S0-BJV-941

A Young ORANGUTAN in a Loving Home

FRANCINE NEAGO AND ESTHER KERR

Uturat Publishing

Chanhassen, Minnesota

Copyright © 2006 Francine Neago and Esther Kerr

All rights reserved. No part of this book may be
reproduced, stored in a retrieval system or transmitted
by any means, whether electronic, mechanical,
photocopying, recording or otherwise without
prior permission of the authors.

Cover photograph: Willis Anderson
Design: Lois Stanfield
Illustrations: Dawn Meader

For information email:
uturatpublishing@fastmail.fm

ISBN: 0-9776819-0-4

Printed and bound in the United States of America

CONTENTS

Malaysia
Singapore
Brunei
Sabah
Sarawak
Malaysia
Sumatra
Borneo
Sulawesi
INDONESIA
Java
Bali

INTRODUCTION

MY NAME IS FRANCINE NEAGO. I am a **Primatologist**, and a fully qualified General Practitioner of Medicine. I also love orangutans. Orangutans are members of the Great Ape family that includes gorillas and chimpanzees. I know how to give medical treatment to these wonderful human-like animals.

Orangutans are native to the islands of Java, Borneo and Sumatra. All these islands are situated in the country of Indonesia that is also part of Southeast Asia. All the orangutans that once roamed Java are now extinct because of hunters; but orangutans still roam the islands of Borneo and Sumatra.

ESTHER KERR is an Australian grandmother, who began writing full-time in 2000 after a varied career. Esther is the author of the non-fiction, juvenile version of *Welcome to Nigeria*, Times Editions © 2002.

Esther met Francine Neago in Malaysia in 2002. Francine asked Esther if she would create children's books from old manuscripts of Francine's adventurous life with animals. Esther agreed because she knows that all animals are Soul just like you and me. There is no better way to show the children of the world that orangutans are Soul, than in story form.

The following story describes a few days in the life of a young orangutan that lived in a loving home—Francine's home. This was on the Island of Java, Indonesia, but the orangutan was born in Borneo. He was the first orangutan that ever lived with Francine.

THE ORANGUTAN
COMES HOME

WHEN MY YOUNG ORANGUTAN first came to live with me, he was only eighteen-months-old. He was a stocky, strange little fellow, and not beautiful by human standards. At first, I thought he looked like a very old, wrinkled, and pot-bellied man who stood about 28 inches (70 centimeters) tall.

He walked upright like a human with a very straight back. His immense feet that looked like a second pair of hands, were turned in. His nose was placed high up in his face, and it spread out wide. His ears looked too small for the rest of his face, and he had a pair of big brown eyes that shone at me with pride and intelligence. Around those big brown eyes, his skin was white and wrinkled. This wrinkled look, combined with a protruding jaw, made him look more like a funny clown.

Rich, rusty-colored hair, about one inch long, covered his body; but his white skin was visible in patches. This gave the appearance that he was wearing a worn-out overcoat. I thought he looked rather ugly, and for that reason, I felt the need to protect him. I agreed to look after him.

I did not have any children of my own, so I was filled with a feverish excitement. It appeared as if a magic wand had been waved to make me a mother immediately, although it was to a young orangutan. Looking down at his intelligent and sensitive face, I wondered what kind of Soul was hiding behind those tender eyes and clown-like face.

The Orangutan's First Meal

My mother and I had just finished our late breakfast. All the dishes were still on the table. I sat this young orangutan on one of the chairs at the table. His chin barely reached the tabletop. I asked my maid to fill the bowl on the table with more fruit. When she returned with the bowl of fruit, she placed it within the orangutan's reach.

The little orangutan observed the bowl of fruit silently for a while before he pointed with his index finger to a piece of fruit. With this gesture he gave a short, soft squeaky sound. He waited a little while before he repeated the same gesture and sound. I handed him a **tangerine**, and he responded with a squeaky sound of a different tone. I guessed that the first squeaky sound was, "Please give me," and the second squeaky sound was, "Thank you."

The orangutan peeled the tangerine in seconds using one hand and his lips. He placed the peels delicately on the tablecloth in front of him. The easy and dainty way he ate amazed me. I was used to how the monkeys that lived with me ate. The monkeys always ran off with the fruit, and ate it while they either hid or swung off my favorite curtains. I realized that an orangutan is as different from a monkey as a lion is from an elephant.

After watching this orangutan's behavior, I understood the meaning of the name *orang utan* that the Dutch, who colonized Java for 350 years, gave to them. *Orang* means *person* and *utan* means *forest*—the person of the forest. (In Sumatra, orangutans are known as *mawas*, and in Borneo they are know as *kalimantans*.)

Half an hour after the orangutan's arrival, he had completely adopted me as a **substitute** mother. He sat on my lap, sought my attention, cuddled and squeezed me, and followed me from room to room with an alarmed cry.

I looked up the name *orangutan* in the dictionary while he helped me to turn the pages—delicately using his lips. I read:

> *Orang utan: an* **anthropoid** *that has the appearance of man. Adult males can weigh over 350 pounds (159 kilograms), females 160 pounds (73 kilograms), reaching maturity at about the age of twelve years. Longest known lifespan in a zoo, 58 years. Long arms, 32 teeth, over-sized body with a shaggy red coat and twelve pairs of ribs. Least known of the great-apes, nearly extinct. Now found only in Borneo and northern Sumatra.*

Not Everyone Is Happy

That first day, my little orangutan learned to shake hands with our maids. I also started to teach him simple words by holding up an object while I repeated its name several times.

When my husband came home for lunch, he demanded that I place the orangutan in a cage. I was equally determined that the orangutan was too young to be put in a cage. In a strong voice, I stated, "He will eat at the table, sleep in his own room, and have the run of the house—just like a child. We have a dog in the house, so why not an orangutan?"

Of course, my husband was angry, and stormed out of the house without his lunch. Needless to say, my husband returned in the afternoon for his evening meal.

The evening meal for the orangutan consisted of rice and bananas with a little warm milk. I knew that his food habits needed to be varied gradually, so as not to disturb his health.

After we had finished eating, I prepared a temporary cot for him. In it I placed a mattress that I covered with two small sheets, and two pieces of woolen cloth. I knew that in the wild, he would have covered his body with leaves. This is done to keep the heat in, and the mosquitoes out. I placed this little human-style bed in an empty guestroom.

When the orangutan saw the bed, he started caressing the woolen blanket. Noticing that I allowed him to touch it, he started pulling it apart while he watched my face for a change in reaction. For ten minutes he turned the mattress over and over, as he beat it with the back of his hand. He folded and unfolded the sheets—holding the edges delicately with his thick, rubbery lips.

Finally satisfied with the mess he made, he lay down—carefully covering every part of his body. I smiled with satisfaction as he fell soundly asleep and began to snore. It was now 9:00 p.m.

The Second Day

The next morning, my lazy little boy, the orangutan, woke up two hours later than the monkeys. It took him another few minutes to be fully alert. Slowly and deliberately, he inspected his room from top to bottom. He acted like someone who wanted to buy the house and its contents. He opened every drawer, sat on every chair a few seconds, and tapped the walls. He also licked and gently sniffed everything he picked up. Each thing he touched was carefully placed back where he found it.

Apparently satisfied, he shifted his attention to the back-yard. There in cages, we kept three monkeys and other animals. He wasn't surprised at seeing them. He acted like he was superior—and he was. In the order of animals, an orangutan belongs to a dominant **species**.

We had Utchi and Utcha, **macaque** monkeys; Klappa, a pig-tailed, macaque monkey; Maw-Maw, a **gibbon;** Gila, a palm **civit**; Goosy, a white goose; many **incubated** chickens; and of course, Doggy.

Throughout the day, the orangutan was quietly sociable to the maids and the cook. If he heard a sound he couldn't readily identify, he threw himself into my arms for protection. He uttered a short, quick, frightened squeak while he pouted his lips, as if sulking. This behavior only lasted a few days.

The Orangutan Is Named

Our maids started calling the orangutan "Tuan" because he behaved like a human. The word *Tuan* means *Sir* in the local Javan language. It was also the way the maids addressed my husband. It was amusing when one of the maids called out, "Tuan." Both Tuan and my husband were uncertain whom the maids were calling, so both answered (with the same grunt) at the same time.

Tuan was like a child who was both dependent and independent. I needed to carry him around a lot just like his mother would have done. I believe that all apes, as well as most animals, usually respond to attention and love. I discovered that my clown-like, ginger-colored orangutan knew exactly how to return love. He had many wonderful ways of showing his feelings. He gave me much more attention and love than I could ever give to him. He was constant in his trust and in his emotions.

SOCIAL ACTIVITIES

ORANGUTANS GROW at approximately the same rate as human children. By the time Tuan was four-years-old, he was almost as tall as a human four-year-old; but his body was much broader. He also had longer hands on the end of long arms, and longer feet on the end of short legs. But, just like a human, he had no tail!

Tuan's strength was also several times that of a human four-year-old. He was capable of shifting an armchair with an adult sitting on it. He could push a cupboard from the wall if he wanted to reach something behind it.

How Tuan Began Each Day

At eight o'clock every morning, my little "boy" opened his large, dreamy and innocent eyes to the big wide world. With sleepy, lazy movements, he stretched each of his limbs in turn. His first thoughts were always on the cook: not because of love for this hardworking woman, but because he knew she would supply him with six or eight bananas for his breakfast.

To eat his bananas, Tuan sat on the floor. He delicately peeled each banana, one by one, as if he were a dainty lady at a tea party. Then he ate each one slowly. Obviously, he had a deep respect for these ripe, golden beauties, called bananas. Following this treat, he drank a large cup of warm milk that had vitamins added. The cook also provided this drink for him.

After a leisurely breakfast that was worthy of King Kong, the famous movie gorilla, Tuan's thoughts always drifted toward me. I was usually lying in bed. Tuan reached my bedroom, by walking through the back door and then through the dining room. When he finally reached my bedroom, he always found the door was closed, but not locked.

To Tuan, a closed door did not keep him from his goal. He was too small to reach the lever type handle, so he carefully placed a chair below the lever, and stood on it. To open the door he leant heavily on the lever. Click! It opened, as if by magic. After the first time he did this, he became very quick at opening my door.

Perched on the chair at the open door, Tuan peeped in to see if I were awake. As there are a lot of mosquitoes in Indonesia, my bed was surrounded by a mosquito-proof net. If he saw no movement through the net, he silently closed the door again, and went away.

A short time later, he returned. Opening the door again, he peered patiently in. Knowing he was there, I kindly moved my body. Immediately, he strode in using a bold and jaunty walk.

With both hands, he separated the net curtain, and pulled open my eyelids with his fingers. Satisfied that I was awake, he hopped into bed beside me. For a few seconds he nestled in my arms; but boys never cuddle for long. With a soft, birdlike squeak, he gave me a loud wet kiss that always melted my heart. Then he left the room so he could play in the morning sun. I looked forward to the next morning to receive this sweet awakening again.

Tuan's First Present

On the day our new Mercedes car arrived, along with it came a new male **chauffeur**. Tuan disliked most men, so he was especially nervous when he saw me drive away with this man. I did not return until lunchtime. All that time, Tuan sat waiting for me in silence, as if he hoped everyone would forget he was there.

Maw-Maw, the gibbon, invited him to play by dancing in front of him, and teasing him with his long arms. Tuan refused to play with Maw-Maw, the monkey or Doggy until I returned.

Having nothing else exciting to do, Maw-Maw, the gibbon, became mischievous. He ran away with a statue that he found on a shelf in the sitting room. This naughty gibbon climbed onto the roof with the statue. Slowly, he examined it before he dropped it—hoping to see it shatter into many pieces. It didn't shatter the first time, so he climbed down, picked it up, and tried again. This was Maw-Maw's favorite game.

When the motor car finally roared into the driveway, Tuan stood up in expectation. He looked undecided whether he would sulk or greet me as if he hadn't missed me. I handed him an interesting parcel, and his face lit up. This was his first present. He unwrapped the present, as if he'd unwrapped presents many times. Inside was a large plastic box on wheels that he could pull along. That night, he insisted on sleeping with it in his arms, although it was too big for the bed.

A Drive In The New Car

In the afternoon, we planned to go for a drive in our new car. We also intended visiting the home of the American **Consul** in Java. After a quick lunch, we were ready to leave. Full of excitement, Tuan climbed into the new car. He inspected the upholstery and checked the springing power of the seats. All was to his liking! When everyone was seated comfortably, the car moved off. Tuan sat on my lap in the front seat while he viewed the driver suspiciously out of the corner of his eye.

The driver, who was an Indonesian man, ignored Tuan. He showed displeasure at driving around the area in a beautiful car with a "monkey" (the name he insisted on calling the orangutan). I think he felt sorry for the master of the house, my husband, who lived with a "crazy white woman" (me). He thought an animal should not be treated like a human.

When the car stopped at the traffic light, people peered into the car. They remarked on the strange appearance of the foreign child. The driver grinned broadly at this—thinking I would be offended.

I was unaware that in Java where I was living, the Javanese people imagined that Tuan was my son. They spread tales about this "orang" of mine. They knew that white-skinned people were very hairy, and white men needed to shave their faces daily. Asian people have few hairs on their faces and chests, so they shave their faces only once per week.

First Social Visit

When we stood at the door of the American Consul's residence, Tuan clung tightly to me. He was a little afraid of the strange surroundings. Regardless, he waited politely with me until the door opened, and we were ushered inside. At my house, Tuan didn't mind being alone. It was his territory, and he felt at home.

On this visit, Tuan behaved very well, and he sat quietly beside me. He accepted the food offered to him, without his usual fuss of sampling. At home, I sampled the food before he ate it, as orangutans become sick easily. If his milk was too cold, or he ate a food to which he was not accustomed, he had a sudden allergic reaction. This often forced him to seek a toilet quickly—not convenient in another person's home.

That afternoon, Tuan had a bowl of milk out of the refrigerator. He drank it in one gulp, and sat chewing on a dry biscuit. After a few seconds, he began to get restless and stood up.

"Wait, Tuan, we will be going home soon," I said, not realizing the serious reason for his behavior.

Tuan sat down obediently. I continued talking to my friend, but soon I felt my sleeve being pulled gently. Again I told him to sit quietly, and wait. He gave me a look of utter despair. A few seconds later, he got up, and dashed to the front door. Pulling a chair over to stand on, he quickly opened the door, and ran outside to the car. I followed closely behind him, but he didn't open the car door. He squatted there, and fertilized the Consul's beautiful garden.

His sensitive body had reacted quickly to the cold milk. The dear little boy had surmounted his strong feeling of insecurity for the sake of hygiene: not messing the floor inside the house.

Tuan's Bath Procedure

From Tuan's first day with me, his daily bath became an important part of his life. Since there were no bathtubs in Indonesia, his bath was prepared with approximately thirty liters of water in a large metal tub.

Before entering the water, he tested the temperature with his little finger. The temperature had to be "just so" or His Royal Highness, Tuan, the Orangutan, might refuse to wash that particular day. He always mimed his displeasure at me if I prepared lukewarm or cold water for him. The kitchen, where the water was heated, was forbidden territory. He did not dare to go there.

17

If Tuan found the bath water to be too hot for his liking, he quickly removed his finger and waved it wildly in the air, as though it had been burned. He cast a look that clearly stated that I was stupid to get the temperature wrong. If I didn't respond quickly enough, he went for a jug of cold water from the bathroom. Without spilling a drop, he added some of the jug's contents to the water already in the tub. After testing the water again, if it were still too hot, he added more cold water until he was satisfied.

When the bath temperature was satisfactory, he climbed clumsily in—first one leg, then the other. From a standing position, he gradually lowered himself until the water reached his chest. Then the fun began!

He started by sampling the soap. If he disapproved of the taste, he didn't use it; but if he liked it, he ate some of it. After this, he soaped himself. Tuan enjoyed the soap bubbles that he created on his hairy skin. It was difficult to persuade him to rinse them off. As with a human child, I had to divert his attention to something else.

He loved to splash the water out of the tub as far as possible, by hitting the surface of the water with the flat of his hand. As he didn't splash me intentionally, I often allowed him to empty the tub this way. Even then, he was reluctant sometimes to leave it.

I wrapped him in a large bath towel that was all his own, and I carried him to the sunny porch. He had goose bumps by then, and I knew how deceptive an Orangutan's strong appearance was. I took no chances with Tuan. A simple common cold could turn into **pneumonia**. It would kill him before drugs had a chance to take effect.

When Tuan was older, I allowed him to take his bath alone, and use a towel to dry his body. To dry his back, he held both ends of the towel. Then he did a wiggle dance, as he rubbed the towel up and down over his back. We had no idea how he learned this effective technique. No one at home dried his or her body in this way.

In the jungle, where orangutans usually live, there are no towels. The tribes of Borneo and Sumatra who live near the rivers, don't use towels. They simply put their clothes back on their wet bodies. Sometimes they even bathe while still wearing their clothes.

Center Of Attention

Tuan loved being pampered. If I was called away while he was still in the bathtub, he neglected to rinse himself off. Dripping wet, he left the tub to fetch his towel from the clothesline. Wrapping himself in it, he lay down on the couch holding his hairbrush ready for me to brush him.

For a long time after he was dry, Tuan gave me his arms, legs, and neck to brush. He pointed to each part of his body that he wanted re-brushed. After a short while, I began to relax on the job. He took the brush and continued to brush himself, but not for long. If he saw my hand lying idle, he placed the brush into it. If there was no reaction on my part, he took my fingers and closed them over the brush handle. If I still refused to brush him, he placed his hand over mine, and brushed himself with both our hands. No words were needed because on his face was a look of strong determination that stated, "Brush me!"

This routine usually took place every morning in front of the house. It gained for him a large crowd of admiring neighborhood children. The children gathered at the gate, on the wall, in the trees or wherever they could find a place that kept them out of reach of Doggy. Doggy strongly disapproved of **gate-crashers**.

Tuan loved the commotion he created amongst the children. He played up to them by using "His Royal Highness" attitude. His open-mouthed audience stared in fascination. I must have appeared to be the slave and obedient servant to His Majesty, the Orangutan. When the children clapped him, Tuan acted as if it were his divine right to be clapped.

Meal Times For Tuan

Orangutans are vegetarians and fruit eaters. If he were free in his natural environment, Tuan would feed on a variety of fruits, the young green **shoots of plants**, leaves, and other wild vegetation. Occasionally, he would eat an egg from a bird's nest, as well as honey, nuts and flowers.

Apart from breakfast and afternoon tea, Tuan's two main

meals were taken at the table with my mother and me. Tuan refused to eat with my husband. My husband felt the same way about Tuan.

To reach the table, Tuan sat on a chair with a cushion on it. It seemed natural for him to eat with us—using a plate, spoon and fork, and drinking from a glass without breaking it. He behaved very well at the table. He pointed to special foods that he liked. He never reached over to take anything from the serving dish by himself.

Tuan's meals were twice as big as the average human meal. We gave him soup, vegetables, and fruit. For dessert, his favorite foods were egg custard, and any fruit that was served—even acidic types of fruit such as oranges or lemons. He always finished his meal with two or three cups of warm milk. Occasionally, I supplied honey and nuts, as well as flowers and plant shoots to keep him healthy.

Some foods we thought were not good for him. We wouldn't allow him to eat potatoes or other starchy foods, as he was already too fat. But the naughty boy stole sweet potatoes from our monkeys when he had a chance. Meat, fish or seafood gave him a skin allergy, and an upset stomach for days.

At afternoon tea, Tuan ate two thick slices of whole-wheat bread with butter and lots of jam on top. Before eating this, he licked the jam off the bread. Smiling sweetly at me, he returned the slice to obtain more jam. If I refused, he reluctantly dipped it into a large bowl of warm milk that had multi-vitamins, minerals, and B-Group vitamins added.

I gave him 500milligrams of B-Group vitamins each day.

This was many times the dose required by a human. If he didn't get this large dose of the vitamin, his skin developed the appearance of dried **parchment**—scaly and whitened.

The Orangutans who lived in the zoological gardens looked scaly and whitened. They didn't get the variety of young leaves and plant shoots that contain the B-Group vitamins in large quantities. I tried this remedy on the Orangutans in our local zoo, and their skin became as soft and supple as it should be.

If I had allowed Tuan to choose his food at each meal, he would have eaten nothing but cakes and milk, so I regulated his meals carefully. When Tuan was with us for one year, we had a cake baked for his birthday as a special treat. Since we didn't know the exact date of his birth, we celebrated the day he became partially humanized.

ATTITUDE TOWARD HUMANS

AN ORANGUTAN WALKING UPRIGHT, choosing his own food in the market is not a familiar sight—even in Java. I soon became used to heads turning, and hearing strange remarks about my "retarded" child who refused to answer when he was asked a question.

"Why won't he talk to me?" was a repeated question. This happened also whenever he sat eating with a knife and fork in a restaurant.

The Indonesian market place had more than twenty varieties of bananas for sale. But Tuan, the **aristocrat**, only liked two sub-species of these bananas. If the banana he was given was not one that he liked, he ate a piece, and then in disgust, he spat it out, as if it were poisonous.

Tuan adored the fruit called **durian**. It is considered a special fruit in Indonesia and Malaysia. The durian contains alcohol, and was very expensive for us to buy. The native people of Indonesia say that it tastes like heaven, but smells like hell. Foreigners always disliked its dreadful smell at first. They said it smells worse than the smelliest European cheese.

Local people can tell how long a foreigner has resided in Indonesia by the person's reaction to the durian fruit. They watch to see if the person is at the *smelling* stage, *tasting* stage, or has reached the *loving it* stage.

Another fruit that Tuan loved was **mangosteen**. If we didn't supply either durian or mangosteen, he unhappily ate mangoes, papaya, or **rambutan**. These fruits all grow naturally in his native jungle.

An Unpleasant Meal Time

One day, my professor husband, arrived home late. He was always bad-tempered when he was hungry. We had already started eating dinner. That evening, my husband decided that he wouldn't allow the orangutan to eat at the table with the "superior beings"—the humans.

"I'm locking the orangutan out of the dining room," he announced loudly.

"If you lock the door, the maids won't be able to enter with the food," I said timidly.

"Then I shall tie him up!" yelled the professor.

"He is smarter than you," I replied, as I became more daring. "He will untie your knots."

"You women don't even know how to make decent knots," he replied haughtily. "I learned to tie knots with the Boy Scouts. It would take even you half-an-hour to untie my knots."

Although Tuan had not finished his meal, our docile and obedient orangutan allowed the master of the house to take him into the garden. I didn't dare interfere. In the garden, Tuan was tied up to an iron railing, as if he were a horse.

Tuan seemed to be perfectly indifferent to what was happening. He didn't even watch from the corner of his eyes. He appeared resigned to his fate, and we felt sorry for him.

"There," stated my husband when he returned to the dinner table. "He'll *never* undo the knot I just tied."

Seconds later, my husband's face turned red, and he became silent. There was Tuan quietly walking toward his vacant place at the table. But Tuan didn't sit down. Instead, he dragged his chair away from the table to a corner of the room, and hopped onto it there. Dangling from his hand was the cord that had been used to tie him up.

Tuan and I exchanged glances. I detected a little twinkle in his eye. He had outsmarted the human male, and he was well aware of the significance of this achievement.

The maid handed Tuan his plate so that he could finish eating his meal. The rest of the meal was finished by all of us in silence. Now it appeared that the orangutan considered it beneath his dignity to eat at the same table as *that man!* Dignity, pride, and contempt were written all over Tuan's clever face.

Behavior With Visitors

When Tuan settled down to home life, he behaved differently with each new person he met. It was impossible to tell beforehand how he would act. Once he behaved in a certain way with a person, he rarely changed.

Sometimes, he was friendly with visitors, shy with others, or behaved like a clown showing off. With certain people, he behaved cold and analytical. But to those who were worthy of his friendship, he was obedient, tender, cuddly, and considerate.

Often he liked to be the center of attention when visitors were present, and he acted as a very demanding child does. If he didn't get enough attention, he threw a tantrum. I gave him his milk in a bottle when guests were offered a cup of coffee or tea. This made him furious because I refused to give him a cup of coffee or tea. He didn't behave like this when he ate at the table with the family.

Ladies

Usually, Tuan liked all ladies, regardless of age. By one glance at a lady, Tuan decided if liked her. Sometimes he became overly friendly with a lady. He sat on her lap, caressed her arm, and looked into her eyes for a reaction. He also showed one or two tricks to make her laugh. With other ladies, he sat on *my* lap, and was demanding before he went away to play alone.

Men

Tuan totally disliked all men, and avoided them. A man probably hurt him when he was younger. If a man tried to hold him against his wish, Tuan would beat the man with the flat of his hand. If this didn't work, he used his strong arms to get out of the embrace. Although some orangutans are known to bite people, Tuan never did that to anyone—even while playing.

Children

Tuan loved children, particularly children his own size. With children, no introduction was necessary. He brought his toys to share without being asked. He had many things that he used as toys. These were objects that he had chosen from around the house, and he treasured them.

His most precious possession was a tin can full of nails, nuts, bolts, and a rubber washer for the water tap. He also had old elastics, strings, magazines (to look at the pretty pictures), odd rags, and a lady's handbag. The elastic served as chewing gum. The washer was used to grind his teeth on. He wove the strings into each other or tied them into knots.

If his toys didn't interest the children, he turned into a clown, danced monkey fashion, and rattled the tin can of nails. If the children didn't laugh enough, he twisted his double-jointed limbs into the craziest positions, and threw himself backward toward them.

Another funny action was to use one hand to curl his lower lip over his chin, and the other hand to pull his upper lip over his flat nose. It was fun to see young children trying to imitate him unsuccessfully.

Making noises

Accompanying clown-ish acts, Tuan made a series of blowing and sucking noises. For adults, Tuan never showed off in this way. For adults, he reproduced a lion's roar that he learned well while visiting the zoo. Sometimes he made the puff-puff sound of the old-fashioned steam engine that was still used in Indonesia.

Other times, Tuan embarrassed us by producing certain sounds that were suggestive and realistic imitations of toilet noises. This shocked some of our guests who didn't know where the noises came from. They didn't expect an ape to make such sounds. The source of the noises became known soon because Tuan repeated the sounds several times. After showing-off in this way, Tuan always ended with vigorous hand clapping, as a wide grin uncovered his teeth—all perfectly white and regular.

New Maids With Bad Attitudes

Adult orangutans hate anyone who mistreats or harms them. Later on, even any person who resembles the hurtful one is disliked. If an orangutan has a bad experience with someone wearing a colored uniform, he will hate everyone wearing that type of uniform. This happens with animal keepers in the zoo.

A new maid came to work for us. After her second day, Tuan had a grudge against her. If she came close to him, he tried to hit her. This was unusual behavior for him. He never acted this way with anyone before. Intrigued, I asked the cook if she knew why he did this. She told me that the new maid tried to burn Tuan's finger with a cigarette lighter. Life became so unbearable that I had to send the maid away.

Another time, a different maid—also new—hit Tuan on the head with a broom handle. Every time he saw the maid with a broom, he backed into a corner, uttered a fright call, and covered his head with both hands. I also sent this maid away.

A CREATIVE AND INTELLIGENT BEING

Tuan's Creative Activities

Threading Leaves

Tuan enjoyed climbing the trees in our small garden. When he was up a tree, he often picked some leaves for a creative activity. To choose the leaves, he picked one, smelt it, and carefully decided if it were suitable to use. When he had gathered a lot of leaves, he tucked them between his toes on both feet to carry them.

After he climbed back down the tree, he found a shady corner of the garden to sit. Feeling totally contented, he spent the next hour threading the stem of each leaf through the next one. He often made colored patterns by threading a pale green leaf through a dark green leaf or through a red leaf.

When he was happy creating his patterns, he ignored everybody and everything around him. I respected his need for privacy at that time. If he had been in his native jungle, he would have spent more and more time away from his mother. This would be in preparation for his solitary wanderings when he grew to be an adult. Apart from meeting the odd female who accepted him, he would most likely spend years and years in isolation from his own species.

Tuan's threaded leaves turned into decorative **garlands**. He was always happy with the results of his work. He adorned himself with these garlands of leaves. Sometimes he made a belt for his fat waist. This required a lot of fitting. He had to keep adding leaves, then trying the belt on; then adding more leaves until it was large enough.

Other times, he fashioned necklaces by twisting the strands of leaves several times around his neck in the **oriental** manner. He even placed a large handkerchief over his head, and tied it under his chin. When he did this, he resembled an elderly Russian peasant of **Mongolian** origin. After the handkerchief was secure, he fitted a crown of leaves on top of his head. Sometimes, he found a cloth to wrap around his shoulders. Can you imagine how he looked?

In whatever attire he chose that day, he walked proudly out of his hiding place. He paraded in front of us so we could admire him, and tell him how lovely he looked. The proud look on his face changed into a full grin that exposed both sets of his teeth. I was sincere in congratulating him on his achievement. Orangutans are very sensitive to compliments. He could never have an experience like that in a zoo.

I wondered how this young "animal" had such a wonderful sense of color and beauty. Where would he have learned such fashions? Perhaps he found the idea in the magazines that he looked through. He often sat cross-legged on the floor staring at the bright colored pages of a magazine. He called my attention if he wanted to share some of the pictures

with me. He held open the page and pointed at the particular thing of interest.

Painting and Drawing

The orangutan was also a talented artist. To make his work of art, he used talcum powder and face powder that he took from the bathroom.

Sometimes, like the painters of olden times, he made his own powder from bits of brick he found in the yard. If he couldn't find a brush, he used his finger, a nail, or even a fork. Any material served as a means of self-expression, just like modern art students discovered. One of his favorite ways of expressing his art was by spreading toothpaste on the bathroom mirror.

Realizing he enjoyed creating in this way, I gave him a pencil and lots of paper. Then he didn't need to find anything to use or to draw on when he was feeling artistic. He took his artwork very seriously now that he had a real pencil and paper like we used.

Depending on his mood, he sat either cross-legged on the floor or with feet wide apart when he drew pictures. Before he started to draw, he flattened the already flat piece of paper. The big decision of how to draw took a lot of time. He had to select which side of the paper to use, and which angle was the best to place the paper on the floor. When he finally decided, he held the paper down on the floor with both his feet and one elbow.

When he picked up the pencil, he carefully examined it, smelt it, bit it, and licked it. Usually, he used his left hand to draw, but not always. After each line or dot, he wet the pencil with his lips and tongue. Sometimes he drew a straight line followed by another straight line across the first one. Other times he drew one dark line followed by a lightly drawn line. He added curved lines, circles and dots in various places over the straight lines. This activity required intense concentration by our great orangutan artist.

Tuan disliked being disturbed or watched while he was drawing. Occasionally, the maid came into the room while he was absorbed in his artwork. He immediately picked up the paper, and held it against his abdomen with both hands, so she couldn't see what he had drawn. When she left the room, he stood up, and with a violent action, slammed the door shut. My presence was the only one he tolerated during this time.

His work of art was usually completed within three minutes. He carried it to me in triumph by holding the paper open on both palms of his hands. Naturally, I lavished praise on him, and cuddled him for a minute.

These drawings were far from stupid. They had a distinctive character. I allowed him to draw at any time he chose without any encouragement from me. Sometimes, he sat to draw, and appeared to lack inspiration. Other times, he decided that his drawing was just scribble, and totally unsuitable to his own artistic genius. He crumpled the sheet of paper, and threw it away in disgust. Five or six drawings might be crumpled before he was finally satisfied.

Instinctive Behavior

Tuan's natural orangutan instinct emerged at the strangest times, but his instinct guided him. He **always** obeyed this inner knowing. One day when the sky was a beautiful blue, without any trace of a cloud, he sensed something. Suddenly, he stopped in the middle of a game, and went into the garden. There, he plucked a large banana leaf from our banana tree. This is always done by the local people of Borneo where he came from, to protect themselves against rain.

He brought the large banana leaf into the house. Then he gathered a few rags and the toys he loved, and placed these treasures in a corner of the living room. There he made his nest by overturning a bamboo chair, and covering it with a sheet to create a tent. By the time the anticipated rain started falling, he was hidden safely underneath the tent.

Of course, he was happy in his hiding place, as he played with the toys he had collected thoughtfully before he hid there. He waited patiently underneath for two  hours until the last drop of rain fell. He knew he was protected on all sides. Only when he was sure the rain had finished, did he come out of his hiding place.

Did Tuan sense that catching a cold from the rain could be fatal to a young Orangutan? Or had his mother taught him this when he was a baby in the jungle? I don't know the answer. But, Tuan loved water, and happily took a bath. He also liked using the garden hose. He held the hose to help the maids clean the animal cages. Yet, he always avoided getting the water sprayed on him. If he became wet by mistake, he used the first towel nearby to dry himself. Sometimes, this was my skirt if I stood too close to him.

Other Skills Achieved

Learning for Tuan was a challenge. He wanted to achieve success in whatever he tried. If he didn't succeed when he attempted something for the first time, he kept trying to do it until he succeeded. This was only if no one forced him into doing it.

A good example of learning a new skill came after he tied knots in my husband's shoelaces. These knots continued along the shoelace until there was no more room. This was definitely not a sensible behavior, especially when my professor husband was late leaving the house. The knots may have been a beautiful weaving pattern to Tuan, but not to my husband. If my husband wanted to wear those particular shoes, he had to untie all the knots first.

In an attempt to prevent Tuan from touching my husband's shoes, I supplied him with strings and laces. But this idea did not work. He preferred to work directly on a pair of shoes. I either had to teach him to tie bows with shoelaces or lock up the shoes.

Tuan and I practiced together, and he was fascinated. For many days after that, he tried tying the bows until he succeeded. It was a proud moment when he brought the shoes to me with the shoelaces tied into bows. After that day, he never tied knots in the shoelaces.

Tuan also loved to teach. He patiently tried to teach Doggy how to do something. When he discovered that Doggy could neither clap hands nor throw a ball, he turned his teaching talent to Maw-Maw, the gibbon. The gibbon became an expert at both throwing a ball and clapping hands. When the gibbon was shown how to tie things, he could only manage simple knots—not bows.

Simple jigsaw puzzles that Tuan was able to put together baffled the gibbon. Eventually, Tuan assumed that Doggy and the gibbon were not very clever, so he gave up teaching them.

Tuan learned to imitate easily. Everything he did was always for fun. I never gave him a food reward like they do in a circus.

My clever orangutan loved to help me nail the lids on wooden boxes that I needed to mail somewhere. I showed him where to put the nail, and he hammered the nail straight in without missing. Sometimes, he hit two or three nails into the same spot. He was very proud that I allowed him to use a real hammer. But this activity didn't last long, because it was too **monotonous** for him.

Later in the day, when he became bored with his toys, he remembered the hammer and nails; but couldn't find them. He decided to create substitute tools. He tested the bottom of a tin with his hands, and decided it could be used as a hammer. He wound off a piece of wire from a cage to use as a nail. As he had nothing to nail, he tried to nail the wire into the concrete outside.

At the time, I was having an afternoon nap The sound of hammering woke me. Fascinated, I watched him from behind the partially closed shutters of my bedroom.

The concrete resisted his nail, so he checked the nail, and then the tin. Satisfied that they were not the problem, he rubbed and licked the surface of the cement. Then he tried to nail into the cement again. Finally, he succeeded in nailing into a crack in the concrete. A look of triumph filled his eyes, and he smiled broadly. Knowing he was safe and happy, I went back to bed.

Other Interesting Behavior

This wonderful orangutan was always active. He thought there were many interesting things to do in a house. These included turning doorknobs; swinging on doors; and lying in the car pretending he couldn't hear anyone calling him.

Tuan had the ability to find several solutions for any problem. To open something, he knew that he could use a lever, a spoon handle, a piece of wood, or a wire bent at an angle. He also used any of these items for a hook or for an extension of his own arm. They helped him to get things that were out of his reach or where his hand wouldn't fit.

Tuan was very curious. Anything he saw for the first time was interesting to him. All things were examined, smelled, tested, and turned many times. After this thorough examination, he decided whether the thing was worth playing with or whether it should be left alone.

If the thing were not alive, he would carefully pull it apart. This was not for the joy of destroying something. It was a serious activity to see how the thing was made or what it looked like inside. He always tried to put it back together again—usually not in the correct way.

Woven articles such as mats, **rattan** baskets, woolen blankets or plain woven material were pulled apart. Of course, he attempted to weave them back to their original state. I convinced him that the doormats were banned from this activity. He quickly understood the doormats needed to be left alone if they were to be useful. With these mats he invented a new game. He shook

the mat to be sure that there was no dust in it. Then he hid underneath to scare anyone passing him.

Tuan created many new games. Boxes and tins were fitted into each other. Sometimes he wrapped them up in a cloth before putting them away carefully. He also took out all the paper from the wastepaper basket. He flattened each sheet, stacked the sheets neatly, and returned them to the wastepaper basket.

The bottom of a plastic water bucket became a drum. Sometimes he placed the bucket over his head, and beat it as he walked along. This was usually to entertain others to gain admiration.

Getting Into Mischief

Even when we were away from home, Tuan, the young orang-utan, was allowed to roam freely around the house. When I arrived home one day, I noticed that the cupboard containing the biscuit tin was unlocked, and biscuit crumbs lay everywhere. When I checked inside the cupboard, the biscuit tin was still there, and the lid was on tight. But, when I checked inside the biscuit tin, I discovered his mischievous activity. The biscuit tin was nearly empty.

He had found my bunch of keys, and had opened the locked cupboard easily. I scolded him with a pat on the back, but I spoke gently. If I didn't do it gently this way, he would have been upset for hours afterwards. Naturally, he refused his next meal because he was too full.

Tuan visited the refrigerator daily. He always made sure that no one was nearby before he entered the room cautiously. To open the refrigerator door (an old lever type), he placed his weight on the heavy handle. After a moment's hesitation, his naughty fingers would dip into the food containers.

He sniffed the food first before putting his finger into his mouth. If he disliked the taste, he wiped his hands on his protruding stomach or on the clean curtains. If he enjoyed the taste, he sucked his finger before dipping it into the food container again. He kept on doing this until he was contented. Naturally, by the time he finished his taste-testing session, he became sick from too much cold food.

If I arrived on the scene suddenly, there was not enough time for him to close the refrigerator door. In that situation, he

looked innocently away in the opposite direction while he swung one foot.

He hoped I thought he was only playing with the open door. Of course, I did not think that, and I promptly threw away any food he had touched. Orangutans have diseases that can pass to humans, and probably a few jungle diseases that we don't know about.

Serious Mischief

The most mischievous thing he ever did happened when most of the household was having an afternoon nap. All the animals were relaxing out of the hot sun, but Tuan felt the urge of freedom run through his veins.

Utchi and Utcha, the monkeys, were grooming each other. Doggy had one eye open as he slept. Klappa, the monkey lazily scratched his head, and Maw-Maw, the gibbon, dozed peacefully in her playpen.

The mischievous orangutan decided to wake Maw-Maw for some fun. He removed a nail from the railing of Maw-Maw's cage. With great patience, he loosened the cage wire by unwinding it in reverse. Finally, he made a hole large enough for his friend, Maw-Maw, to climb out.

As soon as the gibbon was free, he bounced along through the house like a whirlwind. He snatched a bunch of bananas, and broke the dish they were in. His friend, Tuan, opened the double doors to the backyard, and Maw-Maw escaped quickly outside. Happily free, Maw-Maw swung and leapt through the five trees in the yard while Tuan contentedly ate the stolen bananas. After the feast, the two friends hugged each other before they looked for other amusements.

Tuan was a very naughty orangutan because he broke the rules he had been taught. He took his friend to a forbidden room. Inside this room, safely behind a locked door, we kept brushes, paint pots, hammer, and bottles of all sizes. With his great memory, Tuan knew where he had last seen the important bunch of keys. From then on, it was child's play to find the right key to open the door to this forbidden room.

Both of these apes, played for at least one hour with the different powders that we used for antique furniture touch-ups. Maw-Maw decided that his gray wooly fur also needed a touch-up. He chose red powder to do this. Tuan obviously wanted to keep his natural color because he touched it up with brown water paint. I suspected that the brown paint reminded him of his favorite dessert-chocolate. I was grateful that he had not used the oil paint because it would have killed him. He probably did not like the smell of the turpentine in the oil paint.

Not content to just paint themselves, they went to admire themselves in the sitting room mirror. While there, they brushed the excess paint onto the floor. Their awful behavior did not end there. Maw-Maw, like all gibbons, could not stay in the same place for long. He climbed on the curtains, sofa, and the chairs. A trail of powder was everywhere. My sitting room resembled a battlefield with traces of red blood (the red powder) all over it. They must have been very happy because the pattern on their bodies showed signs that they had hugged each other. Tuan's chocolate-colored paint was swirled amongst Maw-Maw's red powder.

By the time I arrived at this scene of horror, Tuan was still busy refreshing the blue carpet with brown overtones. He held the brush as if he were a professional painter. Maw-Maw tried to induce me to join in the fun. My angry voice scared him so much that he jumped into my arms to ask for forgiveness. When I looked out into the backyard, I noticed that the goat had not escaped Tuan's painting skill either. She had brown tones on her white hair.

Because my anger was very strong, I meted out no punishment to the culprits. I was far too afraid these sensitive animals would die from the experience. To wash Tuan was easy, although three rinses were necessary to remove all trace of the paint. He sat obediently on a stool in the bathroom while I threw cold water over him. Then I soaped him before I threw more water on him. This was the only way the paint would come off.

The gibbon was a problem. This type of ape hates water, and his thick fur takes hours to dry. I thought he would catch a cold. I experimented with talcum powder as a dry shampoo. Maw-Maw then became both red and white. Finally, I had no choice but to dip him into water that was the temperature of Tuan's bath. The gibbon became so angry that he bit me in several places. That was the first time he had bitten in his life.

The poor thing looked ugly when he was wet. I nursed him wrapped up in a big towel. To dry his fur quickly, I stood him near the hot oven. He enjoyed this warm air, as he was trembling with the cold.

Next day, both friends showed no sign of getting a cold from their washing experience. The only thing that I noticed was that they both had loose stools when they did their toilet. Other than that, they were none the worse for their mischievous escapade.

From that day on, there were no unhappy incidents, and life continued as usual for Tuan, the orangutan.

GLOSSARY

anthropoid: human-like.

aristocrat: a member of the highest class in society.

chauffeur: a person employed as a driver

civit: perfume, with a strong musk-like smell

consul: officer who represents a government in a foreign country.

durian: spiny fruit with a creamy pulp and foul smell.

garlands: a ring of flowers or leaves.

gate-crashers: uninvited guests at a party.

gibbon: a small, tailless, long-armed ape of Southeast Asia.

habitat: the natural home of a plant or animal.

incubated: hatched eggs.

macaque: a type of Asian monkey.

mangosteen: a white juicy-pulped fruit with a thick reddish-brown rind.

Mongolian: a native of Mongolia, north of China.

monotonous: not varied.

oriental: characteristic of the Far East like Japan.

parchment: sheep or goat skin prepared for writing.

pneumonia: lung disease.

primatologist: one who studies primates: apes, etc

rambutan: a red plum-sized prickly fruit.

rattan: East Indian climbing palm with long thin jointed pliable stems.

shoots of plants: young growth of plants.

species: things with shared characteristics.

substitute: a thing acting in place of another.

tangerine: small sweet orange (origin Tangiers).

MORE READING

Great Apes. Nature Watch series. Barbara Taylor (Anness Publishing, Ltd)

Orangutans. A true book series. Patricia Fink A. Martin (Children's Press)

Orangutans. Animals of the Rainforest series. Christy Steele (Raintree Publishing)

Orangutans. Our Wild World series. Deborah F Dennard, John F McGee (T&N Children's Publishing)

Orangutans. Zoobooks series. (Wildlife Education, Limited)

The Orangutan: forest acrobat. Christine Sourd, (Sagebrush Education Resources)

VIDEOS

Adventures in Asia. Really wild animals series. National Geographic Society. (Columbia Tristar Home Video)

Animal life cycles. Animal life in action series. Burrud Productions, Inc. (Schlessinger science library)

Orangutans, the high society Wild discovery series. Mark Linfield (Discovery Channel Pictures in Asia with Green Umbrella Ltd)

Orangutans. Washington Park Zoo. Animal profile series. Jan Smith. (Rainbow Educational Video)

WEB SITES

www.enchantedlearning.com

www.sandiegozoo.org/zoo/ex_absolutely_apes.html

www.yahooligans.com

INDEX

AFTERWORD

NATURE USUALLY GIVES each of its creatures the necessary skills for survival. Man and the Great Apes seem to be an exception because we both need to be taught. In an Orangutan's natural environment, the mother teaches her young the necessary survival skills. It takes seven long years before the young male or female gain independence.

The Orangutan population decreased rapidly from the illegal poaching, loss of habitat to palm oil plantations, illegal logging, and disastrous fires. Today, the number of Orangutans is slowly increasing because of the dedicated people who work tirelessly to save these loving members of the Great Ape family. The situation is still critical. Much needs to be done.

Francine Neago, who has passed the age when most people retire from work, is more active than ever. She still lives in Southeast Asia, focusing all her energy in one direction only—the survival of all endangered animals, especially the orangutans

Now that you have read this book, I am sure that you understand why orangutans should not be allowed to suffer unnecessarily. Maybe you can be of service to the animal kingdom, as you grow in love and kindness to all life.